APOPHIS ASTEROID

understanding the asteroid
with the biggest impact on
asteroid Exploration

KRISTIE FLEMING

TABLE OF CONTENT

Introduction

The cosmos has long held secrets, and in the dance of celestial bodies, Earth occasionally finds itself in the crosshairs of potential cosmic calamities. One such enigma, named Apophis, captivates our attention with its colossal presence and the rare opportunity it presents for scientific exploration. This chapter embarks on a journey to understand the essence of Apophis, exploring its significance and placing it within the historical tapestry of asteroid tracking and impact scenarios.

Overview of Apophis and Its Significance

Apophis, a celestial wanderer discovered in 2004, emerges as a colossal protagonist in the cosmic drama. This asteroid, a colossal 1,100 feet wide, is set to grace Earth with its proximity on April 13, 2029. A behemoth wider than three football fields, Apophis commands attention not only for its size but for the unique scientific prospects it brings. As it hurtles through space, this cosmic traveler will approach Earth closer than any other object of its magnitude in recorded history.

The significance of Apophis extends beyond its impending celestial visitation. Once considered a potential harbinger of catastrophe, scientists have alleviated anxieties by confirming that Apophis will not collide with Earth for at least a century. This celestial proximity, rather than being a harbinger of doom, offers a rare and invaluable scientific opportunity. It beckons astronomers and planetary scientists to unravel the mysteries hidden within its stony exterior, providing insights into the very fabric of our cosmic origins.

Historical Context of Asteroid Tracking and Impact Scenarios

To truly appreciate the weight of Apophis' significance, we must delve into the annals of history, where the shadows of celestial bodies have occasionally cast ominous silhouettes upon Earth's timeline. The awareness of near-Earth objects and the potential for catastrophic collisions has evolved over centuries, from ancient civilizations peering into the night sky to modern-day astronomical advancements.

The historical context of asteroid tracking is replete with moments of awe and trepidation. Throughout human history, comets and meteors were often perceived as celestial omens, their appearance linked to auspicious or calamitous events. However, it wasn't until the development of telescopes and the systematic study of the night sky that humanity gained the ability to track and predict the trajectories of these celestial wanderers.

The late 20th century marked a turning point in our cosmic vigilance. The advent of space exploration and technological advancements allowed scientists to extend their gaze beyond the confines of Earth. NASA's establishment of the Near-Earth Object Program in the late 1990s exemplified humanity's commitment to understanding and monitoring potential threats from space.

Apophis, with its initial discovery in 2004, thrust itself into the forefront of this celestial surveillance. Calculations revealed a disconcerting 3% chance of a collision with Earth in 2029, sending ripples of concern through the scientific community. Its name, borrowed from the Egyptian god of chaos, reflected the initial apprehension surrounding its potential impact. This unexpected and rare event prompted astronomers to reevaluate their understanding of asteroid behavior and risk assessment.

As the scientific community grappled with the uncertainties surrounding Apophis, the incident prompted a reevaluation of our cosmic preparedness. The Torino scale, a color-coded warning system adopted in 1999, became a crucial tool for assessing the potential danger posed by asteroids and comets within the next 100 years. Apophis, marked as a level 4 on this scale, underscored the seriousness of the situation, prompting concerted efforts to refine its predicted trajectory and assess the true risk it posed to our planet.

In the chapters that follow, we will embark on a deeper exploration of Apophis, unraveling the scientific endeavors undertaken to study this colossal asteroid as it gracefully graces the cosmic stage in 2029. The narrative will weave through the twists of celestial calculations, the relief of revised trajectories, and the unparalleled scientific opportunities that accompany this once-in-a-lifetime celestial event. Apophis, once a potential harbinger of catastrophe, now stands as a celestial muse inviting humanity to peer into the cosmic abyss and glean insights into our own cosmic origins.

Discovery and Early Concerns

In the vast expanse of the cosmos, where the dance of celestial bodies unfolds in sublime choreography, a celestial interloper made its presence known in the year 2004. This chapter embarks on a journey to uncover the story of Apophis, from its humble discovery to the initial calculations that sent ripples of concern through the scientific community. We delve into the naming rituals of the cosmos and unravel the significance of this celestial wanderer that briefly held the ominous potential to alter Earth's destiny.

Apophis's Discovery in 2004

The narrative of Apophis begins with the vigilant gaze of astronomers scanning the night sky. In June 2004, a team of astronomers at the Kitt Peak National Observatory in Arizona, led by Roy A. Tucker, David J. Tholen, and Fabrizio Bernardi, glimpsed an intriguing celestial visitor during their routine sky survey. This newfound wanderer across the cosmic canvas was subsequently designated 2004 MN4.

As astronomers continued to observe and track this celestial vagabond, its true identity began to unfold. Further observations from different vantage points around the globe allowed scientists to calculate its trajectory and parameters, revealing a celestial body with dimensions that demanded attention. The colossal size of the asteroid, estimated to be around 1,100 feet wide, immediately set it apart from the multitude of celestial objects routinely observed.

The discovery of Apophis, named after the ancient Egyptian god of chaos, brought an unexpected and significant addition to the catalog of near-Earth objects. The naming of celestial bodies often carries cultural and historical weight, and in choosing the name

Apophis, astronomers acknowledged the potential chaos this celestial wanderer could sow. Little did they know that the journey of Apophis would become a saga of scientific curiosity and cosmic exploration.

Initial Calculations Suggesting a Potential Impact in 2029

As the scientific community continued to scrutinize the newfound celestial companion, a moment of disquiet unfolded. Initial orbital calculations suggested a disconcerting possibility – a 3% chance of Apophis colliding with Earth in 2029. This revelation set the astronomical community abuzz with a mixture of fascination and concern. The potential impact would herald a momentous event, marking the first instance in recorded history where an asteroid of such magnitude posed a tangible threat to our planet.

The scientific method, with its checks and balances, thrived under the weight of this celestial challenge. Astronomers huddled over their calculations, refining the orbital parameters and revisiting the data to understand the true nature of this cosmic threat. The implications of a potential impact resonated beyond the scientific community, capturing the imagination of a global audience attuned to the mysteries and perils of the cosmos.

In the ensuing months, as Apophis continued its celestial sojourn, astronomers tirelessly refined their calculations. The gravity of the situation was not lost on the scientific community, as the Torino scale, a color-coded warning system for assessing the impact hazard associated with near-Earth objects, marked Apophis with a level 4 rating. This was a rarity; since the scale's adoption in 1999, no other known near-Earth object had reached this level of concern.

The initial calculations, albeit alarming, showcased the precision and rigor of modern astronomical techniques. Observatories worldwide focused their instruments on Apophis, capturing crucial data that would shape our understanding of this celestial interloper and its potential rendezvous with Earth.

Naming and Significance of the Asteroid

Names carry stories, and in the cosmic lexicon, each celestial body's name is a chapter in the unfolding narrative of the universe. Apophis, christened after the Egyptian god of chaos, held a moniker pregnant with symbolism. The naming process for celestial bodies involves a balance between tradition, cultural resonance, and scientific significance.

The choice of Apophis reflected not only the potential chaos the asteroid could unleash but also the reverence astronomers hold for the cultural tapestry woven by ancient civilizations. It invoked the awe-inspiring and sometimes fearsome elements embedded in the mythologies that accompanied humanity's attempts to comprehend the cosmos.

As Apophis journeyed through the cosmos, its name became a beacon of both trepidation and scientific inquiry. The significance of this celestial wanderer extended beyond its potential for chaos; it became a touchstone for humanity's collective understanding of our place in the grand tapestry of the universe. Apophis, once a distant speck in the vastness of space, now bore a name that echoed through the corridors of both astronomical observatories and human imagination.

Astronomical

Precautions

In the vast celestial expanse where Earth dances among the stars, the potential for cosmic collisions looms as a lingering threat. This chapter ventures into the meticulous measures undertaken by the astronomical community to safeguard our planet, with a particular focus on the unfolding drama surrounding Apophis. We delve into the color-coded warnings of the Torino scale, navigate the corridors of NASA's Near-Earth

Object Program, and explore the intricate dance of calculations that led to an evolving understanding of Apophis's celestial path.

The Torino Scale and Assessing Asteroid Danger

In the cosmic theater, where asteroids traverse their elliptical orbits, the Torino scale emerges as a critical instrument for gauging the potential danger posed by these celestial wanderers. Adopted in 1999, the Torino scale provides a color-coded numerical rating that communicates the level of threat posed by a near-Earth object. Ranging from 0 to 10, each level signifies a specific degree of concern, with 0 indicating no likelihood of collision and 10 representing a certain impact capable of causing global consequences.

As Apophis emerged on the celestial stage, the Torino scale became a beacon of caution and awareness. Marked with a level 4 rating, Apophis signaled a situation that warranted attention. This was not a routine cosmic visitor; it was a celestial behemoth that, according to initial calculations, held a 3% chance of colliding with Earth in 2029.

The adoption of the Torino scale signaled a paradigm shift in humanity's approach to cosmic threats. No longer relegated to the realm of speculative doomsday scenarios, the scale offered a systematic and transparent way to communicate the potential risks associated with near-Earth objects. It became a tool not only for scientists but also for policymakers and the public to comprehend the nuanced landscape of celestial hazards.

As Apophis danced through the cosmic ballet, the Torino scale served as a dynamic guide, adapting to the evolving understanding of its trajectory. Astronomers meticulously revised their calculations, fine-tuning the parameters that dictated the potential danger Apophis posed to our planet. The Torino scale, in its color-coded simplicity, provided a visual narrative of the changing perceptions surrounding this celestial enigma.

NASA's Near-Earth Object Program

Amidst the backdrop of celestial alarms, NASA's Near-Earth Object Program emerged as the vanguard of our cosmic defenses. Established in the late 1990s, this program epitomizes humanity's commitment to systematic surveillance and understanding of celestial bodies that share our cosmic neighborhood.

The Near-Earth Object Program operates as a hub of information, bringing together observations from ground-based telescopes, space-based assets, and international collaborations. Its primary mission is to catalog and monitor near-Earth objects, providing crucial data to assess potential threats and, equally importantly, opportunities for scientific exploration.

Apophis, with its initial discovery in 2004, swiftly found its dossier within the archives of NASA's program. The observational data gathered by telescopes worldwide became a puzzle for scientists at the program to solve. The intricate dance of celestial bodies requires meticulous choreography, and NASA's program orchestrates the surveillance efforts that underpin our understanding of the dynamic cosmic landscape.

As Apophis meandered through the celestial expanses, the Near-Earth Object Program became a central hub for coordinating observations, refining predictions, and disseminating crucial information. The collaboration between astronomers worldwide, facilitated by this program, showcased the collective human endeavor to navigate the intricate tapestry of our cosmic surroundings.

Evolving Understanding of Apophis's Orbit

The cosmos, with its boundless complexity, often reveals its secrets through a patient unraveling of mathematical intricacies. Apophis, with its initial unsettling trajectory, became a celestial puzzle for astronomers to decipher. The evolving understanding of Apophis's orbit showcases the relentless pursuit of accuracy and precision that defines the scientific quest.

Initial calculations, hinting at a potential collision in 2029, set the stage for an intense period of scrutiny. Scientists, armed with observational data and computational prowess, engaged in a cosmic dance of calculations to refine the parameters that governed Apophis's celestial path. The orbit, once a source of concern, transformed into a canvas of scientific exploration.

Radar observations became a pivotal tool in this intricate dance. By bouncing radio waves off Apophis and carefully measuring the return signals, astronomers gained unprecedented insights into its orbit. These observations not only refined our understanding of Apophis's trajectory but also laid the groundwork for future celestial encounters.

As the months and years unfolded, the once-shadowy trajectory of Apophis began to crystallize. The uncertainties that shrouded its cosmic path diminished, replaced by a clearer picture of its journey through the celestial expanse. The meticulous calculations and radar observations conducted by astronomers worldwide, often in collaboration with NASA's Near-Earth Object Program, unraveled the mysteries that once veiled Apophis in uncertainty.

The Close Encounter in 2029

As the cosmic clock inches toward April 13, 2029, the celestial stage is set for an extraordinary event - the close encounter with Apophis, an asteroid wider than three football fields. In this chapter, we embark on a journey to unravel the intricacies of Apophis's trajectory, explore the dynamics of its proximity to Earth, and peer into the toolbox of observational instruments that astronomers will employ to study this celestial visitor during its momentous flyby.

Apophis's Trajectory and Distance from Earth

The cosmic choreography reaches a crescendo as Apophis, the colossal 1,100-foot-wide asteroid, approaches Earth in 2029. The intricacies of its trajectory become a focal point for astronomers and space enthusiasts alike. At its closest point, Apophis is projected to pass a mere 19,000 miles (31,000 kilometers) above Earth's surface. To put this in perspective, it's about one-tenth the distance to the moon, an astonishingly close cosmic shave.

The trajectory of Apophis, once a source of concern due to initial calculations hinting at a potential collision, now becomes a cosmic dance that promises not cataclysm but a unique scientific spectacle. This close encounter is a once-in-a-lifetime event, an opportunity for astronomers to study an asteroid of this size at such proximity, providing insights into its physical properties and composition.

As Apophis hurtles through space, its trajectory will offer observers on Earth a visual spectacle. From parts of Europe and Africa, it will be visible to the naked eye on the night of April 13, 2029. In Los Angeles, experienced stargazers might catch a glimpse with binoculars around 3:30 a.m. It won't manifest as a fiery ball hurtling through the heavens, but rather as a star traversing the night sky, a cosmic ballet unfolding in silence.

Observations and Visibility During the Flyby

The night of April 13, 2029, will be etched in the annals of astronomical history as Apophis graces our celestial neighborhood. The observational excitement lies not only in the proximity of this colossal asteroid but also in the visibility it offers to ground-based observers.

Astronomers, armed with telescopes and observational tools, will turn their gaze skyward, capturing a celestial ballet that unfolds with the precision of cosmic choreography. The visibility of Apophis will vary based on geographic locations, offering a unique experience to those fortunate enough to witness this cosmic rendezvous.

In parts of Europe and Africa, where the celestial drama is set to unfold most prominently, Apophis will be visible with the naked eye. The cosmic wanderer, appearing as a star traversing the night sky, will captivate the attention of sky gazers, both amateur and professional. In Los Angeles, with its sprawling city lights, those well-versed in celestial observation might catch a glimpse with binoculars during the early hours of April 13.

The visibility of Apophis during the flyby not only engages the scientific community but also invites the public to partake in the marvels of the cosmos. Observatories worldwide, equipped with advanced instruments, will capture high-resolution images and data that contribute to our understanding of this colossal celestial visitor. It is a shared experience that bridges the gap between astronomers and the wider public, inviting collective awe at the wonders of the universe.

Ground-based and Space-based Tools for Study

As Apophis graces Earth with its celestial proximity, astronomers employ a sophisticated arsenal of ground-based and space-based tools to unravel the mysteries hidden within its stony exterior. The close encounter in 2029 is not just a visual spectacle; it is a scientific bonanza that beckons researchers to delve into the physical properties, composition, and origins of this ancient relic from the cosmic vault.

Ground-based observatories, scattered across the globe, become the eyes of the scientific community during Apophis's flyby. Telescopes, ranging from professional observatories to amateur enthusiasts' backyard setups, contribute crucial data to the collective understanding of this celestial visitor. High-resolution imaging, spectroscopy, and radar observations conducted from Earth's surface become pivotal tools in dissecting Apophis's features.

Radar observations, in particular, stand out as a powerful technique in studying near-Earth objects. By bouncing radio waves off Apophis and analyzing the return signals, astronomers gain detailed insights into its surface properties, shape, and rotational characteristics. This technique, akin to echolocation in the cosmic realm, offers a level of precision that enhances our understanding of Apophis's physical makeup.

In addition to ground-based observations, space-based assets add another layer of sophistication to the scientific exploration of Apophis. Space telescopes, unencumbered by atmospheric interference, capture data that complement and augment the insights gained from Earth. These instruments, orbiting our planet or stationed farther afield, provide a unique vantage point for studying the cosmic dance between Earth and Apophis.

NASA's OSIRIS-REx spacecraft, originally tasked with retrieving samples from the asteroid Bennu, undergoes a cosmic pivot to rendezvous with Apophis post-flyby. Renamed OSIRIS-APophis EXplorer (OSIRIS-APEX), this spacecraft maneuvers toward Apophis, eventually reaching an orbit close enough to collect a sample from its surface. This mission represents a crowning achievement in our ability to explore and interact with celestial bodies, turning a close encounter into an opportunity for hands-on scientific investigation.

The close encounter with Apophis in 2029 transcends mere observation; it encapsulates a collaborative endeavor that spans continents, instruments, and the boundary between Earth and space. It is a testament to human curiosity, technological prowess, and the ever-expanding frontier of cosmic exploration. As the night of April 13, 2029, unfolds, astronomers and sky gazers alike will witness a celestial spectacle that not only graces our night sky but also unveils the secrets of a cosmic wander

Scientific

Opportunities

In the realm of cosmic exploration, the arrival of April 13, 2029, marks not just a close encounter with Apophis but an unprecedented opportunity for asteroid science. This chapter delves into the scientific bounty that this celestial rendezvous promises, exploring the role of NASA's OSIRIS-REx mission, the intricacies of its rendezvous with Apophis, and the ambitious quest to unravel the physical properties and composition of this colossal cosmic wanderer.

Unprecedented Chance for Asteroid Science

Apophis, a name that once echoed with potential cosmic peril, now becomes a beacon of scientific opportunity. The proximity of this colossal asteroid in 2029 offers astronomers and scientists a front-row seat to unravel the mysteries hidden within its stony exterior. This celestial rendezvous is not merely a flyby; it's an invitation to delve into the cosmic archives, decoding the secrets of a relic from the early days of our solar system.

The scientific community braces for an event that occurs only once in a few thousand years—an asteroid of such size passing in such proximity. It's a rare cosmic ballet that provides an unprecedented chance to study an object of this magnitude up close. Apophis, shaped like a peanut shell, holds within its ancient rocks and minerals the untold stories of our cosmic origins.

From the vantage point of Earth, this cosmic wanderer will resemble a star traversing the night sky, its colossal form challenging the limits of our imagination. But for scientists, it represents more than a celestial spectacle; it's an opportunity to conduct a cosmic autopsy, dissecting Apophis to understand its composition, structure, and the clues it carries about the early days of our solar system.

OSIRIS-REx Mission and Rendezvous with Apophis

At the forefront of this cosmic investigation is NASA's OSIRIS-REx mission, originally conceived to retrieve samples from the near-Earth asteroid Bennu. However, with the cosmic alignment of Apophis's close encounter in 2029, the mission undergoes a cosmic pivot, transforming into the OSIRIS-APophis EXplorer (OSIRIS-APEX). This mission, a testament to human adaptability and ingenuity, charts a course toward Apophis, aiming not just to observe but to interact and collect samples from its surface.

The OSIRIS-REx spacecraft, equipped with an array of scientific instruments, becomes a cosmic sentinel on a journey to rendezvous with Apophis post-flyby. This ambitious mission showcases the adaptive capabilities of space exploration endeavors. It epitomizes humanity's capacity to seize unexpected opportunities in the cosmic theater, turning a mission originally designed for one asteroid into a multifaceted exploration of two celestial bodies.

The rendezvous with Apophis, planned shortly after its close encounter with Earth, involves intricate orbital maneuvers. OSIRIS-APEX navigates the vastness of space to position itself in close proximity to the colossal asteroid. The spacecraft, now repurposed for this scientific odyssey, will carefully approach Apophis, eventually reaching an orbit that enables it to collect samples from the asteroid's surface.

This maneuver represents a groundbreaking achievement in our ability to interact with celestial bodies. The OSIRIS-APEX mission transforms a celestial flyby into a hands-on exploration, allowing scientists to not only observe but to touch, in a sense, a remnant from the dawn of our solar system. The samples collected from Apophis hold the key to unlocking the geological and compositional mysteries that shroud this ancient cosmic traveler.

Studying Apophis's Physical Properties and Composition

As OSIRIS-APEX embarks on its mission to rendezvous with Apophis, the scientific objectives extend beyond mere observation. The spacecraft aims to become an intimate companion, scrutinizing the asteroid's physical properties and composition with unprecedented detail. This scientific inquiry into the cosmic relic goes beyond the realms of astronomy; it becomes a journey into the geological history of our solar system.

Apophis, with its peanut-shaped structure, presents a unique celestial canvas for scientists to explore. The surface, etched with eons of cosmic interactions, carries the imprints of the solar system's formative years. By studying the asteroid's composition, scientists aim to unravel the elemental building blocks that participated in the grand cosmic ballet that birthed our sun, planets, and eventually, life on Earth.

Radar observations, a powerful tool in the arsenal of asteroid science, contribute to the unraveling of Apophis's mysteries. The echoes of radio waves bouncing off the asteroid provide detailed insights into its surface features, shape, and rotational characteristics. This echolocation in the cosmic realm serves as a virtual window, allowing scientists to peer beneath the surface of Apophis and decipher the geological tapestry it carries.

The samples collected by OSIRIS-APEX add another layer to this cosmic exploration. By bringing pieces of Apophis back to Earth, scientists gain access to a time capsule from the early days of our solar system. The analysis of these samples becomes a meticulous endeavor, akin to decoding the ancient scrolls of the cosmos. It offers a glimpse into the conditions, materials, and processes that shaped the celestial bodies in our cosmic neighborhood.

Planetary Defense and Impact Scenarios

The cosmos, with its celestial ballet, occasionally brings forth a cosmic threat that demands human ingenuity and scientific prowess. This chapter delves into the historical context of asteroid impacts, the potential consequences of such cosmic collisions, and the proactive measures humanity undertakes to defend our planet. NASA's Double Asteroid Redirection Test (DART) mission takes center stage, offering a glimpse into the strategies devised to deflect potential threats and safeguard Earth from the cosmic bullets that traverse the vastness of space.

Historical Asteroid Impacts and Consequences

The Earth, throughout its geological epochs, has borne witness to the fallout of celestial collisions – a cosmic drama etched in the layers of its crust. The scars of ancient impacts narrate a tale of cataclysmic events that shaped the course of evolution on our planet. One of the most infamous chapters in this cosmic saga is the impact that led to the extinction of the dinosaurs, a cataclysmic event triggered by an asteroid measuring approximately 7 miles across.

The Chicxulub asteroid, as it hurtled toward Earth 66 million years ago, unleashed an estimated 420 zettajoules of energy upon impact. The resulting devastation included a heat pulse that vaporized rock, sparking wildfires that engulfed vast regions. A choking cloud of particulate matter created a prolonged impact winter, casting a shadow over the planet and leading to the extinction of 75% of species, including the mighty dinosaurs.

While the Chicxulub impact is an extreme example, it underscores the potential consequences of asteroid collisions. Smaller impacts, more frequent but less cataclysmic, continuously shape Earth's surface. Understanding the historical context of such events becomes paramount in developing strategies to mitigate future threats.

NASA's Double Asteroid Redirection Test (DART) Mission

In the spirit of proactive planetary defense, NASA introduces the Double Asteroid Redirection Test (DART) mission. Conceived as a technological demonstration, DART aims to showcase humanity's ability to alter the trajectory of an asteroid, offering a potential blueprint for planetary defense against future cosmic threats.

Launched in response to the growing awareness of the potential impact hazard posed by near-Earth objects, DART targets a binary asteroid system named Didymos. The smaller of the two components, Dimorphos, becomes the primary focus. DART, equipped with a kinetic impactor, is set on a collision course with Dimorphos, aiming to alter its orbit in a measurable way.

The kinetic impactor, a high-velocity projectile, is not designed to obliterate the asteroid but to nudge it ever so slightly off its trajectory. This subtle deflection, if successful, represents a groundbreaking achievement in our ability to manipulate celestial bodies and avert potential impacts on Earth.

DART operates on the principle that small changes in an asteroid's trajectory, induced by a kinetic impactor, can accumulate over time, steering the celestial body away from a collision course with our planet. The mission represents a leap forward in our understanding of asteroid dynamics and the potential efficacy of deflection strategies.

Strategies for Deflecting Potential Threats

As the understanding of asteroid dynamics deepens, scientists and space agencies explore various strategies to deflect potential threats. While Hollywood often portrays asteroid deflection as a grand, explosive affair, reality calls for subtler, more calculated maneuvers. Among the strategies considered, kinetic impactors like those employed in the DART mission stand out as a promising avenue for planetary defense.

Kinetic impactors operate on the principle of transferring momentum to an asteroid through a high-velocity collision. By altering the velocity of the asteroid, even minutely, the trajectory can be adjusted to avoid a collision with Earth. This strategy, while seemingly straightforward, requires precise calculations, advanced technology, and a deep understanding of the target celestial body's composition and dynamics.

Another proposed method involves the use of a gravity tractor – a spacecraft positioned near an asteroid, utilizing its gravitational pull to exert a gentle but consistent force. This gravitational interaction, over an extended period, can alter the asteroid's trajectory, steering it away from Earth's path. The advantage of this method lies in its non-destructive nature, avoiding the potential fragmentation of the asteroid into smaller, potentially more hazardous pieces.

A more speculative but intriguing concept involves solar sails – large, reflective surfaces that harness the pressure of sunlight to propel a spacecraft. By deploying solar sails on or near an asteroid, scientists hypothesize that the pressure exerted by sunlight could alter the asteroid's trajectory over time. While this method poses significant technical challenges, it exemplifies the imaginative range of strategies considered for planetary defense.

The quest for effective deflection strategies extends beyond individual missions like

DART. International collaboration and concerted efforts are essential in developing a comprehensive planetary defense framework. Organizations like the International Asteroid Warning Network (IAWN) and the Space Mission Planning Advisory Group (SMPAG) work towards a coordinated response to potential impact threats.

The Future of Asteroid Exploration

As Apophis gracefully passes by Earth, leaving scientists and stargazers in awe, the lessons learned from this cosmic encounter pave the way for the future of asteroid exploration. This chapter delves into the insights garnered from Apophis, the ongoing and anticipated asteroid missions that will build upon this knowledge, and the technological advancements propelling the frontier of planetary defense.

Lessons Learned from Apophis

Apophis, once a harbinger of potential cosmic peril, emerges as a cosmic messenger delivering invaluable lessons to the scientific community. The intricate dance between celestial bodies, the nuances of asteroid dynamics, and the ever-present potential for cosmic surprises become apparent as Apophis charts its course through the cosmos.

The importance of long-term asteroid tracking and trajectory prediction comes to the forefront. Apophis's journey from its discovery in 2004 to the meticulous calculations that ruled out potential impacts in 2029, 2036, and 2068 underscores the significance of continuous monitoring. The dance of celestial bodies unfolds over extended timelines, and our ability to foresee their movements is critical for planetary safety.

Moreover, Apophis reinforces the idea that asteroids are not mere celestial bullets hurtling randomly through space. They follow predictable orbits, and early detection allows for proactive measures rather than reactive panic. The successful prediction of Apophis's trajectory provides a blueprint for future efforts in early warning systems and planetary defense strategies.

The collaborative nature of international efforts in asteroid tracking and planetary defense also stands out. Apophis unites scientists, space agencies, and astronomers worldwide in a shared mission to understand and mitigate potential cosmic threats. The International Asteroid Warning Network (IAWN) and the Space Mission Planning Advisory Group (SMPAG) showcase the power of global collaboration in addressing challenges that transcend national boundaries.

Ongoing and Future Asteroid Missions

Buoyed by the success of missions like OSIRIS-REx and lessons gleaned from Apophis, the future of asteroid exploration is teeming with ambitious endeavors. These missions aim not only to expand our understanding of the cosmos but also to fortify our planetary defenses against potential impact hazards.

NASA's Psyche mission, set to launch in the near future, targets the metallic asteroid Psyche, orbiting the Sun between Mars and Jupiter. This unique asteroid, believed to be composed mostly of metallic iron and nickel, offers a window into the core of a protoplanet that might have undergone differentiation. Understanding the composition and structure of asteroids like Psyche contributes to deciphering the early processes that shaped our solar system.

The European Space Agency (ESA) adds its contribution with the Hera mission, a companion to NASA's DART mission. While DART focuses on redirecting the orbit of a binary asteroid system, Hera's role is to survey the aftermath of the kinetic impact. This joint endeavor enhances our comprehension of asteroid deflection strategies and their real-world applications, bringing us one step closer to a planetary defense framework.

Advancements in propulsion technology also open new avenues for asteroid exploration. NASA's Lucy mission exemplifies this, designed to explore Jupiter's Trojan asteroids. Lucy employs solar electric propulsion, a technology that harnesses solar power to generate electrical propulsion, allowing for extended missions and the exploration of multiple asteroids within a single venture.

Advancements in Planetary Defense Technology

The future of asteroid exploration is intrinsically linked with advancements in planetary defense technology. Lessons from Apophis underscore the need for a proactive approach, and ongoing developments aim to enhance our ability to safeguard Earth from potential cosmic threats.

Beyond kinetic impactors like those employed in the DART mission, researchers explore innovative techniques for altering asteroid trajectories. Concepts like laser ablation, where focused laser beams gradually erode the surface of an asteroid, and ion beam deflection, utilizing charged particles to alter the asteroid's course, represent the cutting edge of planetary defense research. These ideas, while in the realm of theoretical exploration, showcase the creative spectrum of strategies considered for deflecting potential threats.

Artificial intelligence (AI) also plays a pivotal role in enhancing our ability to track, predict, and respond to asteroid trajectories. Machine learning algorithms, trained on vast datasets of celestial observations, enable more accurate predictions and early warnings. The marriage of AI with traditional astronomical methods transforms our approach to planetary defense, providing a force multiplier in our cosmic surveillance efforts.

The development of autonomous spacecraft capable of on-the-fly decision-making adds another layer to our planetary defense capabilities. These spacecraft, equipped with AI systems, can adapt their trajectories and missions based on real-time data, enhancing our agility in responding to unforeseen cosmic events. Such advancements mirror the evolving landscape of space exploration, where human ingenuity converges with technological innovation to secure our cosmic neighborhood.

In conclusion, the future of asteroid exploration unfolds as a captivating narrative of scientific curiosity, technological innovation, and a collective commitment to planetary safety. The lessons learned from Apophis guide us toward a future where early detection, international collaboration, and advancements in technology form the pillars of our planetary defense strategy. 8. Conclusion

Conclusion

As the cosmic ballet of Apophis unfolds and scientists glean insights from this celestial encounter, we find ourselves at the concluding chapter of a saga that combines scientific curiosity, technological innovation, and a collective commitment to understanding and safeguarding our cosmic neighborhood. This conclusion serves as a reflective canvas, summarizing key findings, unraveling the implications for our understanding of the solar system, and casting our gaze forward to future asteroid encounters.

Summary of Key Findings

The journey with Apophis has been a testament to the precision of human ingenuity and the collaborative efforts of scientists worldwide. From the initial concerns raised by Apophis's discovery in 2004 to the meticulous calculations that ruled out potential impacts in 2029, 2036, and 2068, key findings emerge as guiding stars in the realm of asteroid exploration and planetary defense.

First and foremost, Apophis reaffirms the predictive power of science. Through decades of meticulous tracking and trajectory calculations, scientists successfully charted Apophis's course through the cosmos, dispelling the initial concerns of a potential impact. This achievement underscores the importance of long-term monitoring, early detection, and international collaboration in mitigating potential cosmic threats.

The collaborative efforts of organizations like the International Asteroid Warning Network (IAWN) and the Space Mission Planning Advisory Group (SMPAG) shine as beacons of global cooperation. Apophis unites scientists and space agencies in a shared mission to understand and safeguard Earth, showcasing the capacity of humanity to collaborate on matters that transcend national boundaries for the greater good.

The success of the OSIRIS-REx mission and the anticipation of future endeavors like

NASA's Psyche and Hera missions highlight the transformative power of space exploration. These missions not only expand our understanding of the cosmos but also contribute to the evolving narrative of planetary defense. Lessons from Apophis become stepping stones for future missions, enriching our knowledge and refining our strategies for encounters with celestial bodies.

Implications for Our Understanding of the Solar System

Apophis, with its close encounter, opens a window into the mysteries of the solar system and our cosmic origins. Shaped like a peanut shell, this asteroid serves as a relic from the earliest days of the solar system, a time when a massive cloud of gas and dust coalesced to form the celestial bodies that populate our cosmic neighborhood.

The asteroid's composition, a blend of nickel, iron, and silicate, echoes the primordial nature of these celestial remnants. As Apophis bears witness to the evolution of the solar system, scientists anticipate unraveling the secrets embedded in its physical properties. The study of asteroids like Apophis becomes a journey into the past, offering glimpses into the processes that gave birth to our planet and its cosmic companions.

Moreover, Apophis contributes to the growing body of knowledge about contact binary asteroids – celestial bodies with a distinctive peanut-shaped structure. The significance of these peculiar formations lies in their potential to hold clues about the formation and evolution of asteroids. By studying Apophis and its counterparts, scientists aim to decode the intricate dance of celestial bodies and the forces that shaped them over billions of years.

The implications for our understanding of the solar system extend beyond the confines of Apophis itself. The success of ground-based and space-based tools in studying this asteroid sets a precedent for future observations of celestial bodies. As technology advances, astronomers and researchers will be equipped with increasingly sophisticated instruments, enabling them to delve deeper into the mysteries of our cosmic surroundings.

Looking Ahead to Future Asteroid Encounters

The cosmic theater continues, and as Apophis gracefully departs, our gaze shifts to the horizon of future asteroid encounters. The lessons learned from this celestial tango guide us in preparing for the next act, where other cosmic players take center stage in the ongoing exploration of the cosmos.

Future encounters with asteroids hold the promise of unraveling more cosmic mysteries. The OSIRIS-REx mission, now OSIRIS-APEX, poised to rendezvous with Apophis in 2029, exemplifies the continuation of our cosmic quest. The spacecraft's mission to collect a sample from Apophis's surface opens new avenues for studying the composition and origins of these celestial wanderers.

The advancements in planetary defense technology, inspired and informed by Apophis and missions like DART, illuminate the path forward. Strategies for deflecting potential threats evolve with each mission, incorporating innovative concepts like laser ablation and ion beam deflection. The ongoing development of autonomous spacecraft equipped with artificial intelligence enhances our agility in responding to unforeseen cosmic events, showcasing the dynamic nature of planetary defense.

As we peer into the future, the anticipation of NASA's Psyche mission and the Hera mission adds to the cosmic tapestry. Psyche's exploration of the metallic asteroid between Mars and Jupiter promises to deepen our understanding of the solar system's early processes, while Hera's role in surveying the aftermath of the DART mission enriches our knowledge of asteroid deflection strategies.

The future of asteroid exploration extends beyond the confines of our own cosmic neighborhood. The evolving landscape of space exploration envisions missions to distant asteroids, Trojan asteroids, and potentially hazardous objects in the outer reaches of the solar system. Each encounter, fueled by human curiosity and technological innovation, contributes to the collective tapestry of our cosmic journey.

In conclusion, Apophis, with its near miss and subsequent scientific exploration, becomes a beacon in our ongoing quest to comprehend the cosmos. The lessons learned, the implications for our understanding of the solar system, and the anticipation of future encounters converge to form a narrative that transcends the boundaries of our pale blue dot. As we navigate the celestial seas, each asteroid encounter becomes a chapter in the cosmic odyssey, inviting humanity to unravel the mysteries that lie beyond the stars.

www.ingramcontent.com/pod-product-compliance
Lightning Source LLC
Chambersburg PA
CBHW070945290526
45795CB00005B/1651

* 9 7 9 8 8 7 8 2 9 3 4 1 9 *